W9-AZM-818

BUTTER MILK & CHEESE from your BACK YARD

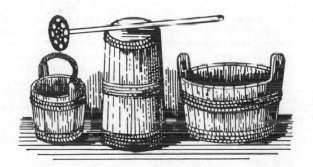

By LEN STREET
& ANDREW SINGER

 STERLING
PUBLISHING CO., INC.   NEW YORK

OTHER BOOKS OF INTEREST

ACKNOWLEDGMENTS

The authors and publisher would like to thank the following for their help in the production of this book: Mr. and Mrs. B. Franche of Martock, Somerset, England, who are champion goatkeepers; Robert and Brenda Vale of Witcham, Cambridge, England, who are ex-goatkeepers and have a house-cow; Mr. R. Wilson of Fairford, Gloucestershire, England, who keeps Dexter cows; and the Department of Animal Science, Cornell University, Ithaca, New York. The following readers of the first edition provided useful information which has been included in this edition: Jim Platts of Willingham, Cambridgeshire, England; Richard and Joan Collins of Westbury, Somerset, England; H. Christian of Woodseaves, Staffordshire, England; and Chris Bennett of Ely, Cambridgeshire, England. Many other readers of the first edition wrote in with useful criticisms and comments, for which many thanks.

Originally published in 1972 by Whole Earth Tools. Second edition published by Prism Press, Dorchester, England © 1975 by Len Street and Andrew Singer. This edition adapted by Anne E. Kallem for American audiences.

Copyright © 1976 by Sterling Publishing Co., Inc.
419 Park Avenue South, New York, N.Y. 10016
Manufactured in the United States of America
All rights reserved
Library of Congress Catalog Card No.: 76-1177
Sterling ISBN 0-8069-3074-8 Trade
3075-6 Library

contents

about

this

book

We have written this book, without apology, as propaganda. Its aim is to provide enough basic information to encourage you, wherever you are, to start your own dairy production. More and more people are getting interested in reducing their reliance on centralized factory food production by growing their own vegetables in a garden; planting fruit trees and bushes; keeping a few chickens, or trying beekeeping. An important element in any move toward increased self-reliance is home dairy production.

This book claims only to be an introduction. It tells you why it is worth starting dairy production; what you will need; what will be involved; and how you can expect to benefit. Once it has done its job—of persuading you to take the first step—its purpose is accomplished. It suggests you move on to more comprehensive or specific works as you get deeper into the subjects concerned.

We hope that you think this book is worthwhile, both now and more especially after you have finished reading it. But even more, we hope that it achieves its stated purpose—to spur you to action—so read on.

Len Street and Andrew Singer

N.Y. Public Library Picture Collection

1 why have a back-yard dairy?

At the very least, home dairy production will cost you something in dollars and cents as well as about half an hour of work per day, every day. It will more probably involve you in quite a lot more. So, why should you bother? We will give you some good reasons.

CASH SAVING

The less cash you need to buy the necessities of life, the less you need to earn by selling your time as an employee, or by selling the products of your work as a self-employed person. So, growing your own vegetables will save you money, and keeping chickens will save you money, but if your consumption patterns are close to Mr. and Mrs. Average Family, a back-yard dairy could save you most of all. If you spend less than average (as we might suspect) on items such as "Meals in Restaurants," "Candies and Sodas" and "Alcoholic Drink," then you probably spend more than average on the products you could produce in a back-yard dairy. At today's prices, a back-yard dairy could provide substantial savings in bought-in food.

ENJOYMENT AND SATISFACTION

For many people, this is the prime motivation in attempting home food production. There is certainly a great deal of pleasure and satisfaction to be had from keeping a house-cow or house-goat. When properly reared, either type of animal becomes an affectionate family pet—a member of the family. Then there is the enormous satisfaction of making cheese and butter. Both demand a high degree of personal judgment and care, but in both cases the home producer can expect to produce products greatly superior to those from the factory. Cheese-making in particular, offers so many variations in technique, and so many different flavors and consistencies that it can become an absorbing hobby in itself. Like wine-making, it has the particular fascination of controlling and utilizing a natural live process.

INDEPENDENCE FROM THE SYSTEM

This may also motivate you to start home dairy production. We can all sit down and moan about the ecological effects of modern farming methods, about the power of modern industrial

N.Y. Public Library Picture Collection

RECEIPT OF MILK AT FACTORY (c. 1875).

corporations and about people having to spend their lives as factory automatons. Unless you do what you can to cut down your economic reliance on this situation, you are as much to blame for what you are moaning about as anyone. Now that you have picked up this book, there is no excuse. We are going to tell you how you can reduce this reliance considerably. Combined with a switch to vegetarian eating, and the cultivation of a small vegetable and fruit garden or allotment, home dairy production could make you 80 to 90 per cent independent of the system in foods.

The milk you are buying right now is probably around three days old; and it has been pasteurized (heated to destroy bacteria). It may also have been left out in the sun and thus had some of its vitamins destroyed. It is probably produced by cows in large commercial herds, pushed by careful feeding of pre-mixed factory foods to their limits of milk production. Treated this way, cows are far more likely to suffer from mastitis and other udder troubles. They have to be injected in the udder with penicillin or

INTERIOR OF A DUTCH DAIRY.

other antibiotics. In a recent study of 40,000 samples of milk, 11 per cent contained penicillin and 1 per cent other antibiotics. This creates a danger of immunity to these drugs in humans and also of possible direct harmful effects on the body. One antibiotic used to treat mastitis, chloramphenical, can fatally disorganize the marrow of the bone. Surely, fresh milk from your own animal *must* be better for you than what you are getting now.

In addition, there may still be danger from the chlorinated hydrocarbon group of pesticides (DDT, etc.) which build up to high concentrations in the fat of animals grazing on treated land and can result in severe contamination of the fat in milk.

Dairy farming, like all branches of farming, has been concentrated into large units by agribusiness. The U.S. now boasts huge "cowtels" which have brought their own pollution problems. One of these super-farms can produce as much cow dung

A NORMAN DAIRY.

as a fair-sized town. And, as you can guess, it is not "economic" to actually use this magnificent natural fertilizer so they have been known to channel it for direct, untreated dumping into nearby rivers, thus throwing out the ecological balance of large river systems.

Apart from the pure health aspects, you may be spurred on faster to home dairy production by a little knowledge of the kind of places that now do the work previously done by hand in cool dairies. Butter is now widely made by a continuous process in large factories, which extrudes it out to be cut and packaged. Some cheese production, notably Swiss and cheddar, has "progressed" to the continuous production line stage and you can be sure that they are working on those that haven't yet. But none go quite as far as processed cheese, defined as:

"The clean, sound, pasteurized product made by cominuting

11

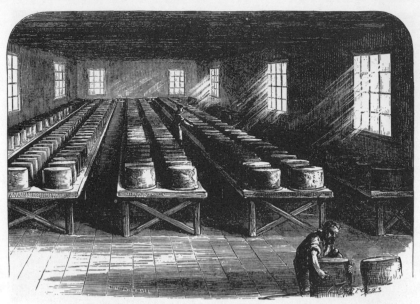

CURING-HOUSE, WHITESBORO CHEESE FACTORY.

and blending, with the aid of heat and water, with or without the addition of salt, one or more lots of cheese into a homogeneous plastic mass."*

Process cheese (made in slices by organizations such as Kraft) is treated with 10 to 20 per cent of a special starter which inhibits certain bacteria naturally present in milk but which causes "gas-blowing" and reduces the keeping qualities of neatly packaged, family-sized cheese portions. The antibiotic used, nisin, is said not to have any harmful effects on the flora of the human stomach. Kraft slices are made by rolling presses and the extruded sheet formed is then sliced into strips, which are fed to the cutting and packing machines.

You may also be interested to know that experiments are underway on irradiating cheese with powerful X-, gamma- and

*"Advances in Cheese Technology," FAO, 1958.

LONGFORD FACTORY (INTERIOR).

other rays to prevent mold growth under plastic wrappers. So, before the dairy technologists get too carried away on their search for the perfectly uniform, infinite shelf-life dairy product, perhaps the time has come to become independent of the whole unpleasant scene.

THE BASIC ECONOMICS

OF HOME DAIRY

PRODUCTION

THE PERFECT MILK PAIL.

If you want to learn more about becoming more independent in dairy products, the first shock we want to get you over is that there is no half-way house. With cereal products, for instance, you can save an enormous proportion of bread costs by buying wheat in bulk and grinding and baking at home—without having to get involved in the actual growing of the grain. The same is not true of dairy products. You need around 10 pints of milk to make one pound of butter. So you can see very quickly that buying retail milk for butter-making is totally uneconomic, certainly at the present subsidized prices we pay for butter.

The picture is not so clear-cut with cheese. You need about 8 pints of milk to make one pound of hard cheese, so that at present prices, it could just be barely worthwhile making your own cheese from bought retail milk. You may feel like attempting to find a source of milk sold cheap because it is slightly turned and so get into cheese-making that way, but the point we want to make is that, generally, home dairy production only really becomes worthwhile when you have your own milking animal.

The following is taken from the United States Department of Agriculture Leaflet No. 538:

"It will pay you to keep a dairy goat if the cost of the milk your family needs is as much as 14 cents a day for the 275 to 300 days a year that a goat produces.

14

If you buy a dairy goat for $50.00
And sell her in 5 years for $10.00

Cost for 5 years is $40.00

Cost for 1 year is . . . . $ 8.00
Interest on $50 costs you . . . . . . $ 2.50
Breeding charge is $10.00
Doe eats about 450 pounds of oats (1½ pounds a
day for 300 days) $12.00
And about 500 pounds of alfalfa hay (2½ pounds
a day for 200 days) $10.00
And root crops . . . . $ 5.50

Which adds up to $48.00
But you sell a kid for $10.00

So, keeping a goat for 1 year costs $38.00
. . . or about 10 cents a day."

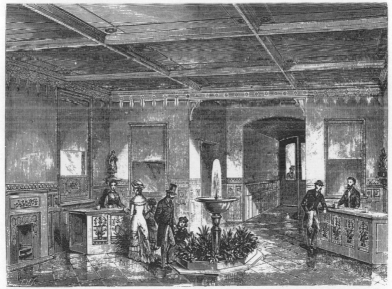

THE SHOP OF HOLLAND PARK DAIRY.

Now, these are 1966 statistics, but the percentages and proportions are easily translated into today's prices.

Two very important factors must be looked into before you consider the purchase of either a goat or a cow. First of all, are there any restrictions in your area concerning the keeping of such animals? Check with your local health department or police department. Even though there might be zoning laws, in some cases all you need is a special permit. Second, if you intend to sell *any* of your milk or milk products, you will undoubtedly have to conform to a state sanitary code. Permits and licenses are available through your local county health department, so contact them for all information you require. Just be sure to do all this before you have your heart set on your own back-yard dairy.

N.Y. *Public Library Picture Collection*
THE RIDGE CREAMERY.

2 what animal to start with

All mammals produce milk, but only certain ones are really suitable as regular milking animals, namely, the larger domesticated breeds, the cow, the horse, the goat and the sheep. Unless you want to go deeply into self-sufficiency and double up your milking mare as a low-cost form of transport, we suggest you forget the horse. Milking breeds of sheep exist in France, where sheeps' milk is used to make such prize cheeses as Roquefort, but if you are not desperate for the taste of back-yard Roquefort, we suggest you forget sheep as well.

This leaves us with cows and goats as the most suitable animals for house-milkers. Both are kept for this purpose and there are many owners with a strong preference for one or the other. This can be confusing for the would-be back-yard dairyman seeking advice, so we will try first to give a reasonably unbiased assessment of the main points in favor of each.

IN FAVOR

OF

GOATS

- Goats are delightful animals to keep—they are full of individuality and character.
- They will eat almost anything. They can survive on land where most cows would starve.
- A goat is a more efficient converter of foodstuffs into milk. For every 140 lb. bag of dairy cake fed, the average goat produces 2 gallons more milk.*
- The milk production of a goat is about right for the dairy product needs of the average family. It needs less space than a cow; costs less to buy; and requires less equipment and less time to look after.
- Goats' milk is said to be easier to digest. People suffering from eczema, asthma and psoriasis are often helped by changing from cows' to goats' milk.

*"Goat Husbandry," David MacKenzie, Transatlantic.

18

- With a small trailer, your house-goat can be taken along on family trips. A cow requires someone at home every day to milk it.
- Goats' milk can be frozen.
- Goats' milk is the closest to human milk and contains 10 times as much iron as cows' milk.
- In most areas, you will be able to get your doe bred so that you do not need to keep a buck.
- Artificial insemination is available in some parts of the country for a moderate fee.
- Goats are far less susceptible to tuberculosis and brucellosis.

IN FAVOR OF COWS

- You are used to the taste of cows' milk. Goats' milk tastes pretty similar cold, but gives off a "goaty" smell when heated. (This is due to the presence of short-chain fatty acids in the milk. These, in ewes' milk, are responsible for the "peppery" taste of Roquefort compared with Stilton.*)
- Most breeds of cow will not destroy young trees and bushes as will goats.
- Surplus bull calves can be sold or raised to produce veal and beef. Goats' meat is not as valuable.

*"Cheese: Volume I—Basic Technology," J. G. Davis, J. and A. Churchill, 1965.

N.Y. Public Library Picture Collection

- A well-reared house-cow is gentle, affectionate and docile. In comparison, most goats are boisterous and capricious.
- If properly trained, a cow can double her usefulness by pulling the plow in your vegetable garden as well.
- Cows'-milk cream settles easily and can be obtained without the use of a mechanical separator. The butter is just as you would expect it to be. Goats'-milk butter can be greasy.
- A cow grazes happily with little trouble to its owner and can be left out day and night, summer and winter. A goat has to have shelter at night and even during the day in case it rains.
- Artificial insemination is universally available from a veterinarian or government service for cows, but not always for goats. Goats often have to be transported to the billy for mating.
- The labor costs per pint of milk are much lower. This, of course, is the reason why virtually all commercial milk production is from cows.

21

N.Y. Public Library Picture Collection

LETTER K COMPOSED BY B. BETTE (1785).

OUR ADVICE

Frankly, our advice is that the animal with which you start will depend upon these circumstances:

1. If you or one of your family suffers from eczema, asthma or psoriasis, try a diet excluding all cow-based products (meat and dairy). If it helps, then buy a goat.

2. If you have less than about an acre of decent grazing, you are going to have to bring large quantities of food, either gathered or bought, to a cow. The economics of home milk production then become marginal, so go for a goat. It eats less and is less fussy than a cow.

3. If your land is of poor quality, it will probably support only a goat or a small, hardy breed of cow such as the Jersey or Dexter Kerry.

In all other circumstances, we recommend a cow rather than a goat as the most suitable house-milker. This is mainly because it is so much less demanding and troublesome, and also gives you 100 per cent acceptable milk and easy cream. A cow is very satisfying and relaxing to keep, but it must be admitted that it is a fairly formidable animal for a complete beginner to manage. Whatever your circumstances, there is a lot to be said for starting off with a goat, even if you intend to progress later to a cow. You can learn how to milk more easily on a goat and you will make less expensive mistakes.

Goats also have one unfair advantage over cows for any beginner. There are more "amateur" goatkeepers around than cowkeepers. They are enthusiastic, they run clubs where you can learn a great deal and they are generally very helpful to beginners. (See Appendix for listing of goat clubs.)

So start with a goat and then, unless your circumstances are as 1, 2, or 3 above, move on later to a cow.

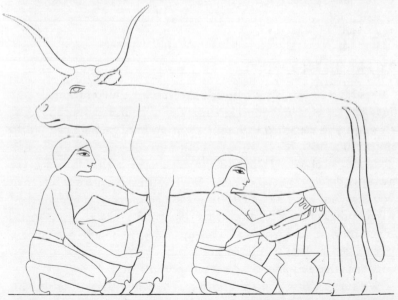

N.Y. Public Library Picture Collection

ANCIENT EGYPTIANS MILKING.

3 making a start

with a goat

We are going to assume that you will follow the advice in the last chapter and start off with a goat, possibly progressing after a couple of years' experience to a cow. If you want to go straight to a cow, read all this selectively and try, if you can, either to start with a calf or get lots of local help—from sympathetic farmers, vets, county farm agents, 4-H clubs, and so on. You will find your State Agricultural Extension Service helpful in supplying detailed information on dairying with goats and cows at home. (See Appendix for address in your state.)

WHAT GOAT TO BUY?

Apart from choosing between the various different breeds, the first choice is whether to go for a non-pedigree "scrub" goat through the classified columns of your local newspaper, or to invest in a pedigree from a reputable breeder. A "scrub" kid or goat may cost very little and the advantage is that if you make awful beginners' mistakes and she dies or goes out of milk, then you have not lost much. The disadvantage is that you get attached to an animal that will more than likely turn out to give an indifferent yield and you will end up doing as much work for 2 pints a day as you would to get 8 pints from an animal in which you had invested and risked a bit more money.

We suggest that you start with a 4- to 6-month-old female kid, *handreared* by a breeder recommended by other goatkeepers. That is the cheapest start, but of course, you have to wait another

24

18 months or so before you get any milk. A year-old pedigree goatling would cost more and save you 6 months of waiting.

There is a great advantage in getting a kid that has been hand-fed from birth. It will respond positively to humans and be much easier to handle.

N.Y. Public Library Picture Collection

As for breeds of goats, there are a number to choose from:

The SAANEN is a Swiss breed, common in the U.S. and Canada. It is pure white and capable of the highest yields on decent grazing.

TOGGENBURG, another Swiss breed, is brown with white markings. It is smaller than the Saanen and less high-yielding, but it is the best breed for grazing rather than browsing. It is the most common milk goat in America.

ANGLO-NUBIAN is heavier than the Swiss breeds and it has floppy ears and a Roman nose. It comes in various colors and gives the creamiest milk, but not in great quantity. The breed is very noisy and nervous.

FRENCH ALPINE is an American milk breed, developed from imports in the 1920's. It is either black or white, or combinations of these two colors.

ROCK ALPINE is an American milk goat. Large and vari-colored, it is the result of mixing early French Alpines with other Alpine breeds.

Any of these breeds is suitable for the beginner, as long as you recognize its problems; for instance, the nervousness of the Anglo-Nubian. In certain circumstances, the fact that goats prefer browsing on a variety of plants, shrubs and trees may be a disadvantage. The little Toggenburg is the best breed to adapt to a life of grazing on grass and little else.

Having said all that about breeds, the actual breed you will choose almost certainly will turn out to be less important a factor determining success or failure than the milk-yielding qualities of the particular stock from which your goat was bred. Yields vary enormously, from under 2 pints a day to 2 gallons a day and more. Our advice is that, unless you are prepared to treat it as a short-term learning exercise only, pay more and get an animal whose pedigree promises a good milk yield.

Judging the likely milking quality of a kid is no matter for a beginner. That is why we emphasize getting advice. The best indicator, if you know how to judge it, is the kid's pedigree, but

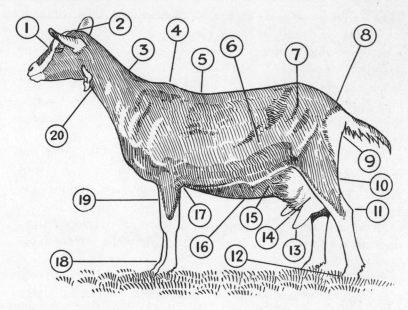

GOOD POINTS IN A MILK AND BREEDING GOAT: 1. EYE, BRIGHT AND GENTLE; 2. HEAD, SHAPELY AND INTELLIGENT; 3. NECK, LONG, NOT COARSE; 4. SHOULDERS, CLEAN AND NEAT; 5. BACK, THE LINE LONG AND LEVEL; 6. RIBS, DEEP AND WELL SPRUNG; 7. PELVIS, WIDE; 8. RUMP, SLOPING GENTLY; 9. ESCUTCHEON, WIDE AND REACHING HIGH; 10. REAR OF UDDER, WELL DEVELOPED; 11. HOCKS, WIDE APART AND STRAIGHT; 12. FEET, SOUND AND NEAT; 13. TEATS, POINTED AND DIRECTED FORWARD; 14. UDDER, SPHERICAL AND FIRMLY ATTACHED; 15. BARREL, AMPLE FOR FOOD; 16. MILK VEINS, PROMINENT; 17. BODY, DEEP, ALLOWING ROOM FOR HEART; 18. PASTERNS, FAIRLY STRAIGHT; 19. FORELEGS, STRAIGHT, SOUND, NOT TOO CLOSE; 20. THROAT, CLEAN AND FINE.

clearly you want a bright, healthy looking kid with an alert eye. The best goats for milking tend to be non-stop nervous eaters. Lastly, unless you want your goat for chasing unwelcome guests out of your yard, make sure that the kid is hornless or has been dehorned (called "disbudding").

HOUSING AND MANAGEMENT SYSTEMS

Depending upon your land and your preferences, there are various systems of management suitable for a house-goat.

1. *Extensive*

If you are surrounded by open scrubland or meadows where a roaming goat can do little damage, you can let it do just that, providing that there is somewhere for it to shelter when it rains and at night. Your goat will gather a lot of its own food, but reckon on providing concentrates all the year-round and bulk food in the winter. (For an explanation of "bulk foods" and "concentrates" see page 38 on feeding.)

N.Y. Public Library Picture Collection
SHELTER COVERED WITH STRAW (1893).

2. *Fenced*

If you have one-third acre or more of decent grazing, your goat will make the best use of it if you fence it into at least two sections, ideally more. Use the sections in rotation. Depending on the acreage, you may have enough grass to make winter hay or silage,

which otherwise will have to be brought in or substituted for. On a very small acreage, extra bulk food may also be needed in the summer. Concentrates will be needed all the year. Most breeds of goats like a variety of bulk foods, so you may have to bring in twigs, tree-leaves, garden waste, and so on, for them. That is why the Toggenburg is so well suited to this system—it is generally happy with just grass.

3. Zero Grazing

This new system, labor-saving and excellent if you are out all day, produces big milk yields, but you should not keep a single goat this way unless it is in sight of company a lot of the time. The idea is to provide housing with feed racks and a straw-covered sleeping area together with an outside concrete run attached. All food is brought into the goat, so it wastes no energy in gathering and very little in keeping warm. Much more is therefore converted into milk. Surprisingly, goats thrive under this system and produce splendid yields. It does not sound very labor-saving, but proponents claim that it is easier to mechanically cut and gather a row of grass every day than it is to maintain and move electric fences in such a way as to make equally efficient use of limited grazing space.

CONFINING

YOUR

GOAT

The goat is an inquisitive animal, fond of eating all sorts of plants you would rather she did not—like prize rosebuds about to bloom, or your neighbor's tomato plants—so she usually needs to be confined. You can either tether or fence her. The trouble with tethering is the labor involved. You have to move her each day to a piece of ground soft enough to take the tethering stake and hard enough so that she will not pull it out. Then when it rains you have to take her in, or move a little portable shelter each day as well. In our opinion, tethering, though widely practiced, is a poor way of confining goats.

The real way is to fence. Non-electric fences have to be 4 feet high to contain a goat effectively, with posts every 6 feet. They should be constructed of sheep fencing, or chain link, but not barbed wire. Now this kind of fencing has, of late, become enormously expensive, even if you construct it yourself. So we would only recommend it for zero-grazing exercise yards or for permanent fencing near the goat shed. For other situations, electric fencing is the cheapest solution and probably the most secure. Its only disadvantage is the labor involved in maintaining it.

An electric fence for goats consists of three strands—one at goat-eye height; one just below goat-belly height, and one half-way between the other two. It is cheaper to make your own posts and buy just the insulators, but follow the instructions of the fencing power-unit manufacturers. Most units go "blip" every second regardless, but modern electronics have now made available a unit that only goes "blip" when the goat touches—clearly much, much more economical to run and about the same price.

Goats are stubborn and determined animals once they decide they would like to get outside their field. To get your goat to respect the electric fence, you will have to wait until it is beyond the playful kid stage. Then get a neighbor or friend to hold its mouth around the fence wire for three or four good jolts. Unless it is a very stubborn goat, it will not go within a foot of the wire again—ever. Nor will it go near the person who held it—that is why it is not a job to do yourself.

The problem with electric fences is that if the bottom strand touches wet grass, it grounds and wastes electricity continuously. Cows conveniently reach under the fence and munch as far as they are able to reach without getting a shock, but goats are too careful to risk that kind of unpleasantness. So you have to go around with a pair of garden shears every so often cutting back potential grounding grass. Some power-units flash lights when grounding is happening, so if you check every day, you need only to cut back the grass when it is actually causing a short circuit.

HOUSING

YOUR

GOAT

N.Y. Public Library Picture Collection

SHELTER OF POLES AND BOARDS (1893).

If you are using zero grazing, you will need a large goathouse with room for the animal to wander around—a minimum of 100 square feet. Otherwise, if the house is merely to be used as a rain shelter and sleeping quarters a goat can manage with something a quarter the size, but it is convenient to have a shed big enough to do your milking in as well, and to store your feeds (but in a 100 per cent goat-proof part of the shed because overeating can easily kill your goat), so we would advise something about 5 feet by 10 feet at the smallest. If the shed is wood, it should be double-walled with some form of insulation between. Storing hay above the goat area is both convenient and insulating for the goat. The shed should be light and well ventilated.

The floor is easiest if made of concrete. If you lay a dampcourse (a damp-resisting layer) below the concrete, and lay a thin top layer of strong concrete directly over a 2-inch layer of polystyrene insulation, you can get away without straw bedding. The concrete floor is then warm enough for the goat to lie down on directly. If the floor is made to slope gently to a gully at one wall, which itself drains outside the shed into either a septic tank or a regularly emptied garden manure tank, then the floor can be swept clean every morning with little trouble. Mucking out (removing the manure) straw bedding is a job you do not have to do often, but it is a messy business. Even if you do use straw, though, the slope-plus-gully design helps to keep the straw dry for longer.

The housing should include provision for feeding concentrates and bulk feed and drinking water. Buckets for concentrates and water should be held so that they cannot be kicked over or trodden in. Bulk feeds are best fed in wall-mounted feed racks made out of rope netting, wood slats or wide metal wire-mesh, all of which allow the goat to pull down hay in a way that comes naturally to a tree-browsing animal.

When designing your feed racks, there is one detail you should allow for. Goats are fussy; they ignore even the tastiest food once it has fallen on the floor, yet having no hands to manipulate hay and such, they inevitably drop a sizeable proportion of what they pull down to eat. The answer is to arrange for any food that drops from the rack to fall where it cannot be fouled, and where it can be collected later to be put back up in the racks.

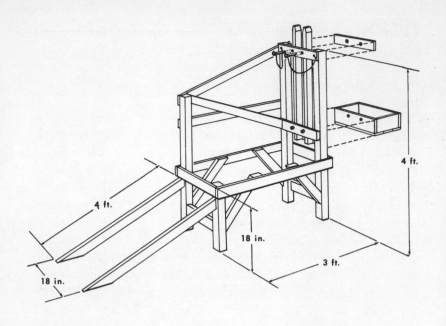

4 ft.

4 ft.

18 in.

18 in.

3 ft.

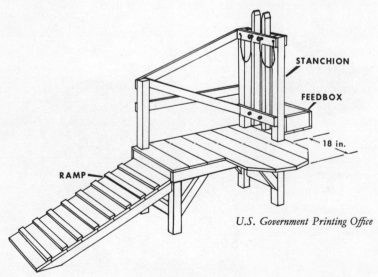

STANCHION

FEEDBOX

18 in.

RAMP

U.S. Government Printing Office

MILKING STAND.

OTHER EQUIPMENT YOU WILL NEED

Goats are rather low animals, so we recommend some sort of milking platform. It should be at easy sitting height and just big enough for the goat to stand on comfortably, but not big enough to let her move around and make milking difficult. There should be some arrangement for holding or tying her head at one end— a movable vertical bar holding her at the neck works well. The alternative is to feed her the concentrates while she is being milked—that will keep her occupied.

The drawings on the page opposite are taken from the United States Department of Agriculture Leaflet No. 538 and show you how, if you are handy, you can build your own milking platform for your goat. The instructions are as follows:

"Dimensions on these drawings can be used as a guide for building a milking stand. The upper drawing shows construction of framework without flooring boards. Bolts at top of stanchion pull out so that uprights can be spread apart to admit the doe's head. Then bolts are put back, and the uprights hold the doe in place. The lower drawing shows the finished stand. For platform, cross supports, feedbox, ramp, and ramp cleats, use 1-inch-thick boards Use 2- by 4-inch lumber for leg braces and ramp frame. Use 4- by 4-inch lumber for leg frame."

Apart from the stand, all you really need is a milking bucket, but it must be stainless steel.

Optional extras include a leather goat collar, a wall-mounted bucket holder, a hoof-paring knife, a coat for when she is sick, and Swiss-type goat bells. For the modern automated goat, you can go in for a milking machine with special goat-sized teat-cups which will cost you a great deal, but that is for later.

Some of this equipment is available through agricultural equipment suppliers. If you have difficulties, consult a goat club (see Appendix).

4 feeding

your goat

We cannot hope to do much more than introduce you to the basic techniques of feeding in this book, but we want to emphasize that, economically, this is the part of back-yard dairying that matters most.

Unless you have enough land for winter and summer bulk food, back-yard dairying is essentially a waste-disposal industry. Putting it another way, if you have too little land yourself to feed the goat, find an unwanted product of someone else's land to supplement it. You are converting things like barley straw, which farmers burn every summer, into rich nutritious milk. The goat is your conversion apparatus. What wastes should you start looking for? Use your Yellow Pages to probe local sources of:

- Pea stalks and pods (cannery waste which can be dried or made into silage).
- Barley straw.
- Bean straw, oat chaff, etc. (less palatable than barley straw, but usable).
- Cornflour residue.
- Brewers' or distillers' grains.
- Sugar-beet pulp.
- Fruit canning and jam-making wastes.

The basic principle of feeding is that a goat uses food for two purposes. Firstly, for the maintenance of her own body and necessary functions and secondly, for the production of milk. The quantities of nutrients required for maintenance depend primarily upon the weight of the goat, though they are, of course, affected by the temperature in which she lives and the amount of walking she does. The quantities required for milk production depend essentially upon the quantity of milk being produced.

Now, in common with other animals, the goat needs a combination of various nutrients for both purposes: starch, protein, vitamins, minerals, and so on, and it would take a computer-controlled feed rationing machine to get them all dead right every day. For simplicity, farmers usually concentrate their efforts on the two most important: starch and protein.

According to Mackenzie, a goat needs for maintenance 0.9 lb. of starch equivalent and 0.09 lb. of digestible protein per day per 100 lb. of bodyweight. For milk production, she needs, in

addition, 3.25 lb. of starch equivalent and 0.5 lb. of digestible protein per day per gallon of milk being given. The starch equivalent and digestible protein contents of various foodstuffs can be found by looking up in tables. Mackenzie's book provides an excellent one for goats.

One method includes the following advice: first you work out how much starch equivalent and digestible protein your goat needs per day and you then estimate what she can probably get from grazing or browsing. On pasture, you should reckon she consumes 20 lb. of grass and 10 lb. of mixed tree leaves, gorse and so on. On rough land, reckon on her gathering 5 to 10 lb. of heather tips, gorse, or whatever, per day. Having then done that calculation, you can work out what starch equivalent and digestible protein you will need to supplement. See which foods give you the required total for the least outlay. Find the cheapest local bulk food available first and then supplement it with a concentrate to make up for low starch or protein.

You can try "challenge feeding." That means gradually increasing the concentrate ration until the milk yield stops increasing. In practice, if you stay a week at a time at each level, you can judge pretty well after a few weeks what the right level is.

Apart from protein and starch, your goat will need minerals and water. Vitamins are not really a problem in goat husbandry. Minerals are usually provided by giving the goat a mineral lick in its shed. These are available from any local feed merchant. The goat takes a lick when it feels a need. Most mineral problems in goats are problems of imbalance or excess. Calcium and phosphorous are the two main mineral needs and should be balanced. Some goatkeepers favor providing licks or boxes of powder for each of the essential minerals separately, on the theory that the goat balances her own diet that way.

What we have said so far seems to suggest keeping the food rations constant throughout the year. That is not the best plan. Nor can you hope, by subtle feed-boosting, to get your goat to defy nature and produce a steady volume of milk throughout the year. You might expect a goat's feed needs to fluctuate in tune with the seasonal availability of foodstuffs. Well, they do and they don't. The trouble is that most births are in February or March, so that in fact the goat in kid needs steadily more and more to eat from November on. After April, her appetite gradually declines again.

POISONOUS PLANTS

You should be aware that the following plants are poisonous to goats and cows: laburnum, rhododendron, yew, beet leaves, buttercup, marsh marigold, Dutchman's-breeches, mountain laurel, poison hemlock, loco weed, buckwheat, and bracken fern. For antidotes, consult your local vet.

Many plants affect the taste of milk, butter and cheese. If you suspect that your milk is tainted, the U.S. Department of Agriculture or your State Extension Service might be of help to you (see Appendix).

5 how
to milk

Milking is fairly simple to master on most cows or goats, although there are individual animals that are very difficult milkers. The principle is to block off the teat, full of milk, from the udder by squeezing with the index finger and thumb, and then to follow this with a squeezing downward and out of the milk trapped in the teat, using the other three fingers. Unless you cut off the milk in the teat from getting back into the udder, no amount of squeezing will get much milk to flow. After practice, the action simplifies into an even squeezing downward from index to little finger. The trouble is that this feels wrong at first. The natural tendency is to want to start squeezing with the little finger. Once you get over this, you should be all right. If not, you may have a difficult animal and therefore you should seek advice.

Here are some more tips:

- Clean a cow's udder before milking, using a swab with warm water. A goat's udder does not usually need cleaning before milking.
- Let the first squirt from each teat go outside the bucket. This avoids using old milk too long in the teat.
- Massage the udder thoroughly to persuade all the milk down.
- Start using both hands. Do not be tempted to learn one hand at a time.
- Do not squeeze above the actual teat. This can damage the tissues of the lower udder.

Although goats and cows produce best if milked twice a day, as near to every 12 hours as possible, the back-yarder can make do with less production, and milk just once a day for convenience. A half-way house to this approach is to use a cow partly as a milker and partly to suckle and rear calves for eating. You take the morning milk, having kept the calf out of reach of the udder overnight, and let the calf take the evening milk.

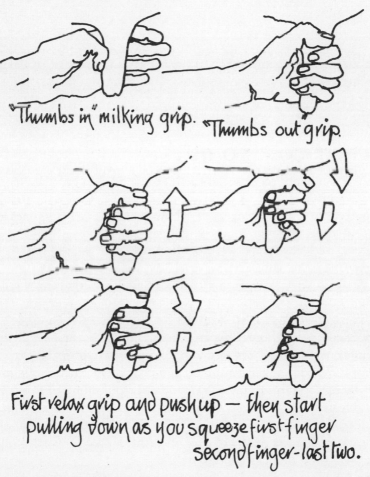

"Thumbs in" milking grip. "Thumbs out grip.

First relax grip and push up — then start pulling down as you squeeze first finger second finger - last two.

6 stage two: progressing to a cow

This section should be read in conjunction with what has already been said with reference to goats. We are assuming that you are already a goatkeeper and are about to move up to Stage Two and obtain a house-cow.

WHAT COW TO BUY?

We would not recommend the inexperienced to get a first cow from a cattle market. A good source of help is the local agricultural vet or county farm agent whom you will also have met as a goatkeeper. He will know a lot about the local dairy herds and, if you are lucky, could put you in touch with someone with a suitable cow to sell.

An idea worth considering is to start with a fairly elderly cow from a commercial herd. She is perhaps no longer giving enough milk for the herdsman to keep her on, but she could be ideal for you. Scout around and see what you can find. If the first vet is not heplful, try another. The vet will help you, too, in making sure that the cow or calf is tuberculin-tested, vaccinated against brucellosis and free of mastitis. (Do not be put off by all this talk of diseases. They are all fairly well under control now in the United States.)

In our experience, there are only three breeds worth considering for a milking house-cow.

The JERSEY is the beautiful doe-eyed cow of cows and the ideal family cow. It is small, manageable and docile and it produces the creamiest milk of all. Because of its size, the Jersey requires less nutrients than other cows and can prosper on smaller, scantier grazing areas. It is as irresistible a pet as a Siamese cat. Its disadvantage is that the bull-calves will not produce saleable meat. There is nothing wrong with the meat but there is not much of it and the fat is yellow. Most people will not buy yellow-fat meat. However, if you are happy to consume all surplus bull-calves yourself, why worry?

The GUERNSEY is very similar but slightly bigger and it gives better yields, although the milk is less creamy. However, the cream content is still above the average. The milk has a characteristic yellow color. The Guernseys are said to be less temperamental than Jerseys, although we have yet to hear of temperament trouble in a Jersey.

The DEXTER KERRY is the mini-cow which was the traditional house-cow of the Irish peasant. It has short legs, is black and

N.Y. Public Library Picture Collection

rather unattractive. On the average it stands 39 inches at the shoulder. It is a very practical little house-cow because it will eat just about anything that a goat will and manage, of course, on less than a full-sized cow. The milk production of the Dexter ranks close to the Jersey's. The meat is first class and produces excellent small joints.

Other common breeds include:

The HOLSTEIN, the largest of all dairy cattle, and thus not particularly suited to back-yard dairying. The black and white Holstein is the highest producer of milk, but does not do well on poor pasture, requiring a substantial area to meet its nutritive requirements.

The BROWN SWISS matures slowly, but has a long life. Only slightly smaller than the Holstein, it ranks second highest in milk production.

The AYRSHIRE has the third highest milk yield and is of medium size.

It all depends upon your circumstances and requirements. Jerseys and Guernseys thrive on good quality pasture, hay and silage. They need their own land. Dexters are more adaptable and can be used, like goats, partly as converters of "wastes" from somebody else's land, although they should have access to fresh grass for at least some of the year. They also give less daunting volumes of milk, particularly in the summer. The trouble is that there are very few Dexters still around. You should not have too much difficulty tracing either Jerseys or Guernseys locally.

INTERIOR OF IMPROVED LONDON COW-SHED.

48

HOUSING AND MANAGEMENT

The three basic systems apply (see page 29), although rough pasture grazing would only really suit the Dexter Kerry.

N.Y. *Public Library Picture Collection*

CONFINING YOUR COW

Cows are less likely than goats to be attracted out of a field by tasty bushes and trees, but, if they are lonely or in heat and can hear other cows, they can be almost as energetic in their attempts to get out. Give a single house-cow plenty of your company and affection and keep her well confined during heat. Then ordinary farm fences and walls should be enough, although electric fencing allows you great flexibility in rotating your available grazing. Cows need less drastic training onto electric fences than goats. Put some tasty food on the other side of the fence while you watch. A few jolts and your cow will get the message.

N.Y. Public Library Picture Collection

PENS AND FRAME OF ARCHWAY FOR A SHELTER (1893).

HOUSING YOUR COW

Again the needs are similar as for goats—just larger. The great advantage, however, is that a cow can happily be left out summer and winter, day and night, if you so wish. Milking can also be done outside, but it is as well to do it under cover in the wetter weather.

FEEDING YOUR COW

The same principles apply as for goats, although Jerseys and Guernseys are much more fussy about "wastes." For these breeds, hay and silage are the only really acceptable bulk feeds.

SHARING

THE LOAD

We have already emphasized what a responsibility it is having either a goat or a cow. Having a cow, you are also faced with having to make use of fairly large quantities of milk. So, what is wrong with owning a cow in partnership with friends? A good scheme we know of is run by two couples, one with two acres and the other in the nearby town. They share the milk and the feed bill half and half but, for their work, the country people get the calf every year as well. The only drawback to this idea is that you must have some method of getting the milk to your partners cheaply at least every other day.

N.Y. Public Library Picture Collection

SEVENTEENTH CENTURY GERMAN DAIRY.

7 dairy products

Milk is a pretty good food as it is. In fact, there are some primitive peoples who live on virtually nothing else. So why bother converting it into other things? The traditional reason must have been to produce from it foods t'... could be kept for the winter, but today this reason hardly applies since adequate supplies of milk are commercially available all year round. No, butter and cheese are still made today because they represent concentrated sources of fats and proteins which are economic to transport over longer distances than milk and because they have their own consistencies and flavors prized independently of milk.

The back-yard dairyman can obtain excellent year-round food from milk alone (assuming he has plenty of winter feed for his animal) but butter and cheese will further enrich his life. Not only that, but instead of throwing the inevitable surplus summer milk to the pigs or chickens, it can be converted into concentrated foods which can be stored.

But dairy work is time-consuming. It has to be done with reasonably fresh milk, of which the home dairyman will have limited quantities, and the trouble is that it takes almost as long to make butter and cheese from 5 gallons of milk as from 50. So, is it worth it? At today's subsidized prices, you get a low return on your labor time for making your own. But they will be superior, after practice, and greatly satisfying. Cheese-making is as absorbing a hobby as home wine-making and, in many ways, very similar.

HOW IS MILK PRODUCED?

Milk is produced by all mammals as a food for their young, during the early part of their lives when their digestive systems are not able to take other foods. It is manufactured in the mammary gland of the mother animal. Production is induced just before the mother gives birth—ready to give the young their important first food. Actually, this first milk is rather special. It is called "colostrum"—it is golden yellow and as thick as double cream. It is designed to clear the inside tracts of the newborn animal. Often a cow or goat will produce more than the infant animal needs. Even diluted one to four with ordinary milk, it will set like an egg custard. The taste of colostrum is an added bonus for the home dairy producer.

Production of milk continues regularly in the mother animal for a period known as the lactation period. This varies considerably in length, but the point is that if you want your animal to continue giving you milk, you must have it mated regularly to make sure that it stays "in milk" for a good proportion of its life. Although both cows and goats have been known to continue a single lactation for three years and longer, most people feel it is safer to get the animal pregnant regularly every year.

An animal will only stay in milk so long as all the milk she produces is being used. This is nature's regulating system. In the normal course of events, the mother would thus only continue giving milk until its young start eating other foods. If you milk the animal dry twice every day, she will continue to produce a good quantity for longer, but you must keep milking dry. This is known as "stripping."

Stripping

The mother can only really be milked if the udder is "let down." Some outside stimulus, such as the cry of its young or the arrival of its owner, induces the production of a hormone, oxytocin, in the pituitary. It takes about one minute in a cow for the oxytocin to reach the udder and cause the muscles to relax to let down the milk. The hormone continues to be sent down for about a further 8 minutes, during which time you should have completed the milking. The important thing is not to stop for breaks. Once you start, you have to carry on until the animal is dry.

The process of milk production in the udder is not yet fully understood, but it appears to be a combination of filtration of certain constituents from the bloodstream, cell degeneration and most important of all, synthesis of other constituents by cell metabolism.

The simplest mammary gland known is that of the Australian duckbill platypus, where droplets of milk ooze out of a small hole onto the animal's hair, where they are licked off by the baby platypus. The udders of the goat and cow are far more developed,

54

with milk tracts leading down from the main producing tissues in the gland, contained in the milk sac, to a reservoir just inside the teat. The teat is a sort of automatic portion dispenser, making sure that the infant only gets down as much milk as he can swallow comfortably. The udder of the cow is divided by thin membranes into four quarters, each with its own teat and each needing to be milked separately. The udder of the goat is divided into two sections.

WHAT IS MILK?

N.Y. Public Library Picture Collection

"Milk is an oil-water emulsion, the continuous phase being aqueous."* That means that it consists of water with certain substances dissolved in it, plus little fat globules floating around in it. The proportions of the various constituents in milk vary considerably, but an average make-up of scientifically analyzable materials is:

87% water
4% fat
5% milk-sugar or lactose
3% casein
½% albumin
—————
99½%

(The last two are the proteins in milk.)

In addition, there are bacteria, many of which arrive in the milk during and after milking, some of which are useful in the human stomach or in the process of cheese-making, but others of which

* "Richmond's Dairy Chemistry," Griffen, 1953.

56

are harmful. There are probably also loads of trace elements in milk, necessary for good nutrition but not easily analyzed out. Milk is known to be rich in calcium and in vitamins A and B. It also contains phosphorus, potassium, iron, and vitamins C and D.

WHAT

IS

CREAM?

N.Y. *Public Library Picture Collection*
CENTRIFUGAL SEPARATOR (c. 1872).

The little fat globules rise to the surface of settled milk and form cream. Eventually, the fat globules will settle at a certain concentration at the top of the milk and the familiar cream line appears. This process can be speeded up in a centrifugal separator. The butterfat content of cream is very variable—between 10 and 65 per cent fat, but 45 to 55 per cent fat is a more normal level.

WHAT IS BUTTER?

Butter is formed by persuading virtually all the fat globues in the milk to join up into a continuous mass, separated from the water and its dissolved constituents in the milk. The persuasion that seems to work best is severe agitation at controlled temperatures, but the exact mechanism by which it occurs is still a matter of dispute.

Butter is made up, very roughly, of:

84% butterfat
7% water
7% unremoved buttermilk
2% salt, added for flavor
———
100%

The "buttermilk," left behind after the butter separates out, is made up of:

90% water
5% milk-sugar
4% milk-proteins
1% butterfat
_____
100%

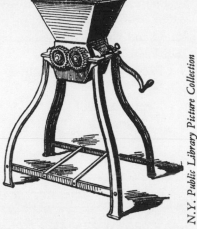

N.Y. Public Library Picture Collection

WHAT IS CHEESE?

CURD-MILL.

If the acidity of milk is increased, the separate particles, particularly of milk protein, coagulate together into a jelly-like or custardlike substance called curd. A great deal of the water and a little of the dissolved substances separate out into a greenish, watery liquid called whey. This separation takes place more easily at above 60° F. The increase in acidity can be achieved by one of three methods.

1. Simply add an edible acid, such as lemon juice or vinegar, to warm milk.

2. Let warm milk stand where it stays warm. Little organisms in the milk (*streptococcus lacticus*) convert the milk sugar, lactose, into lactic acid and presto, the milk curdles by itself. But if you leave it too long, another load of organisms take over and turn it into "bad" milk, so take care.

3. The conversion of lactose to lactic acid can also be achieved by using enzymes. Rennet is the usual one employed. It is taken from the stomachs of calves and processed commercially. Very small amounts of commercial rennet added to warm milk will produce curd.

59

To make cheese, the curd is separated from the whey. It is then eaten fresh as soft cheese, or pressed into molds to dry out the whey and water completely, thereby leaving it in a form which will keep longer.

Cheeses vary enormously in composition but here is a rough guide:*

Percentage of	Camembert	Cheddar
Water	45–51	27–34
Fat	21–30	26–30
Proteins	18–23	27–40
Lactic Acid	0.5	1.5

Cheese is valuable, not merely as a way of storing milk for winter, but in its own right as a concentrated protein food. The other great thing about cheese is that you can make so many delicious variations.

HOOP FOR FLAT CHEESE.

N.Y. *Public Library Picture Collection*

There are plenty of variables in cheese-making:
- the method of increasing the acidity.
- the temperature at which the curd is formed.
- the acidity at which it is formed.
- whether heat is used to produce further separation.
- how much heat and for how long.
- whether the cheese is pressed.
- under what pressure and for how long.
- whether the cheese is affected by other bacteria.

We will have more to say about the use of these variables in re-creating famous traditional cheeses later in the book.

*"Richmond's Dairy Chemistry," Griffen, 1953.

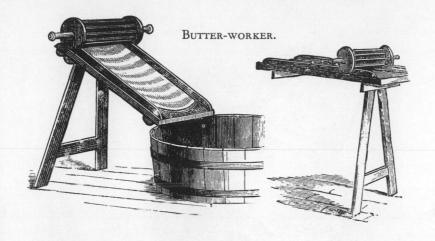

BUTTER-WORKER.

HOW SOME OTHER MILK PRODUCTS ARE MADE

Homogenized Milk is milk, heated to reduce the surface tension of the fat globules, then forced through very small holes at high pressure to reduce the globules to a size at which they no longer rise.

Condensed Milk is milk, mixed with sugar, then heated to around 235° F. and condensed in a vacuum.

Dried Milk is concentrated milk. This is achieved by blowing hot air through the milk, thereby concentrating it. This concentrate is then placed on a rotating drum which pushes out the rest of the water. It becomes brittle solid and is then ground into powder. It is now also produced by a freeze-drying process. Some dried milk is dried whole milk, some is dried skim milk.

Canned Cream contains, on average, half the butterfat content of fresh cream. It is sterilized at 250° F. and should keep for a year in a can.

Casein is sometimes separated out in large dairies. It is used to make artificial milk powder, paper coatings, cold-water paint and glue, and in printing fabrics.

61

N.Y. Public Library Picture Collection

DAIRY TOOLS.

8 producing cream at home

Cows' milk separates into cream and skimmed milk fairly easily under the action of gravity alone. Goats' milk does not and the only effective way of producing fresh cream from goats' milk is by means of a separator. Frankly, this is a bore. But if you are a goat-keeper and sufficiently interested in cream, then try to pick up an old farm hand-separator by asking around at local farms or by advertising. A new model will cost you quite a lot.

Before you buy an old one, make sure that the tinning is unbroken and that all the cones are still there. The separator gives 99 per cent of the available cream and you can adjust it for the thickness of cream you want. The trouble is that even the

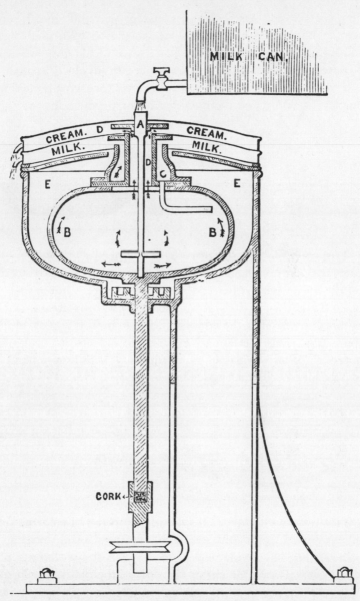

MILK CAN.

CREAM. D A CREAM.

MILK. D MILK.

E C E

B B

CORK

SECTION OF SEPARATOR.

simplest machine has about 23 different parts, all metal, which need to be washed in soda and sterilized after each use. Even if you use a dairy disinfectant rather than boil the components, it is quite a chore when the amount of cream involved is small.

If you are the happy owner of a cow, then cream-making is simple. There are two methods for home production.

MILK-PAN.

N.Y. *Public Library Picture Collection*

THE SETTLING PAN METHOD. You will need a big wide shallow pan (not plastic), and a cream scoop (a saucer will do), to draw off the cream once it has settled. Leave it for 24 hours in a cool place (longer for goats' milk), then skim off the cream, starting by breaking it off at the edges of the pan first. It comes off like a thick skin. This method gives you around 90 per cent of the potential cream available. It needs very little equipment, but quite a bit of cool shelf space.

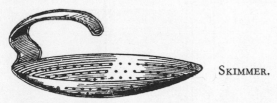

SKIMMER.

N.Y. *Public Library Picture Collection*

THE SIPHON METHOD. (This was developed by Len Street.) For this you need gallon jars, preferably caterers' salad dressing jars because they have wide necks, and a wine siphon from a pharmacy. Slice an oblique cut at the end of the plastic pipe on the siphon. Keep the milk in your gallon jars for 24 hours in a cool place. Then slide the siphon pipe down inside the edge of the jar past the cream layer to the bottom of the skim milk below. Siphon off the skim milk, finishing with the jar tilted over until you reach the visible cream line. You are left with cream in your jar. This method involves less storage space and the equipment is easier to get than a settling pan, but it is a bit more trouble.

Of these methods we favor the siphon. The jars can be placed right in the refrigerator. This speeds settling and gives a better-tasting cream than the settling pan method, because the large surface area of the milk in a pan tends to pick up taints.

If you have a refrigerator, there is no need to make cream every day. The milk will store for three or four days, so you can get away with making cream twice a week if you wish. How much cream you get varies, but it takes about 8 to 10 pints of milk to give 1 pint of cream.

The skim milk can be drunk fresh, but if you have enough whole milk left over for drinking, you won't want to bother with the thinner stuff. It is an excellent food for feeding to calves, kids, chickens, pigs and other animals. Other than that, it is fine for turning into yogurt or skim-milk cheese (see recipes later on).

CLOTTED CREAM

Clotted cream makes a pleasant change for "cream teas" and the like. It keeps longer than fresh cream and so is better for butter-making. It means you can spread your butter-making sessions out to once a week or less. You may find that clotted cream is the only foundation that works for butter-making when the milk starts to get thin in the autumn. If you keep a goat, clotted cream is the only type you can make without a separator.

Stand the milk in a heatproof pan to a depth of 6 to 8 inches, in a cool place for 24 hours. Transfer the pan gingerly to your stove and heat up gently to 150 to 170° F. for use as clotted cream. It will form a ring at the edge of the pan, and the cream will split away. Let the pan stand to cool for 12 hours, and then skim off the cream. It is not a lot of work, but it does take a long time.

WORKING THE BUTTER.

9 butter-making at home

Here is a very simple butter-making method from Jim Platts of Willingham, England:

"We don't own a milking animal. We buy ordinary milk (not Gold Top) and take the cream off each bottle using a 5 ml. plastic syringe. (Your doctor will throw away several of these every day.) We put the cream in a two pound Kilner jar (the wide top makes it easy to get the butter out), and when it is half full, we stand it near the fire until it is just warm to the touch. Then we shake it by hand until pellets of butter form.

We reckon usually on less than five minutes' shaking, and a yield of ½ oz. of butter per pint of milk. We produce as much butter as we want from our weekly milk supply."

Now, for those who want it, here is a little more detail, and some ideas on scaling up the process to use kitchen equipment. There are three types of cream that can be used to make butter.

FRESH CREAM is fine if kept refrigerated prior to butter-making. Alternatively, it can be kept by adding at least 8 per cent by weight of salt, but this produces a very salty butter not to everyone's liking.

CLOTTED CREAM keeps well before the next butter-making session and the taste of the butter obtained is particularly liked once people get used to it.

SOURED CREAM (or Ripened Cream) makes a butter with a distinctive soured flavor. Making butter with soured cream is easier—it turns granular more quickly. For a start, we suggest you try fresh cream. If you experience difficulties, or feel like a bit of variety, try one of the other two.

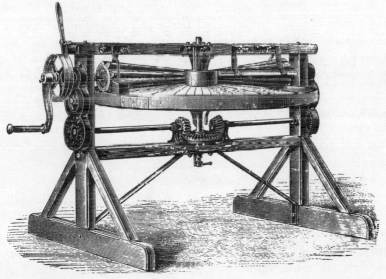

COMPOUND BUTTER-WORKER.

N.Y. Public Library Picture Collection

Keep the cream good and cool before butter-making. There is nothing to beat a refrigerator. Leave at least 24 hours between the last addition of cream and the butter-making session. Stir the cream jar well each time you add a new load of cream.

You do not need to use a special butter churn for butter-making, but they are good for larger quantities. If you do not want to track down a churn, and if you are lucky enough to have an electric food mixer, then you have nothing to worry about. But first get your cream to the right churning temperature:

52 to 60° F. in hot summer weather;

58 to 66° F. in cooler weather.

For this, we suggest you invest in a proper dairy thermometer.

Take a cupful of the thickest cream first and spoon it into the mixer, working at a slow speed (around 90 r.p.m.), with a cake-mixer head fixed. The bowl and head should be cold. Watch until patterns are made, rather like those that appear with whipped cream; then add another cupful of the cream and repeat until all the cream is in. Carry on mixing until the color changes to golden yellow, and then until the wheat-sized granules of butter are formed.

You can then just squeeze the buttermilk out in a muslin bag. This will remove most of it and give you butter suitable for the next week's use, but if you want something you can store, refrigerated, for months, you had better wash it and work it well.

To wash the butter, you will need water at about the same temperature as the buttermilk. Pour some into the granular butter in your bowl and carefully fold the granules over to wash out all the buttermilk. Strain off the wash-water and repeat until it

N.Y. Public Library Picture Collection
SKIMMING BENCH.

stays pretty clear. You will probably need about five washings to get to this stage.

After the final straining out, sprinkle on some salt if you like; we suggest one dessertspoonful per pound of butter. For the really professional touch, add some annatto butter color—particularly if the butter looks less golden yellow than the shop-bought stuff you are used to. It probably will, except in the very best grazing periods of June and July. (See the cheese section for how to make your own coloring.)

If you want the butter to keep, its water content should be under 16 per cent. Commercial butter is more likely to be around 12 per cent. To remove the water, work the butter. The simplest method to start with is to use a cutting board and a strong wide-edged knife or spatula. Squeeze down the butter onto the board, forming a thin layer. If you slope the board (say, on the edge of the drain board), the water you squeeze out will run off immediately. Fold it over and repeat the squeezing down, without rubbing or smearing the butter. Keep it fairly cool during working and keep going until you really are not getting much more out.

Later on, you may feel like buying or constructing something more elaborate in the way of a butter-worker, but that is beyond the scope of this book. Whatever you use, keep it spotlessly clean. It is best to keep a special board for working, rather than the one you used yesterday for crushing the garlic.

BUTTER-MAKING PROBLEMS

There is really only one big problem that occurs in butter-making and that is, for one of many reasons, the butter just does not "come" in 20 to 30 minutes. If 40 minutes have passed and you still have white cream, it could be for one of the following reasons:

- Your cream is too warm or too cold. Sometimes in winter it really needs to go up to 65 to 70° F.
- Your equipment may not be really clean and sterile. This may cause what is known as "ropy fermentation."
- The animal may be almost through her lactation, in which case, butter is harder to make.

Here are a couple of other tips we have heard but not fully tested out. If goat's cream gives you trouble getting too big a granule size, add a cupful of water per $1\frac{1}{4}$ pints of cream as soon as the granules appear. Also, it is said that thinner cream is better in summer and thicker cream is better in winter.

Do not throw away the buttermilk. It contains the milk-sugar and milk-proteins. Drink it fresh, use it for bread-making instead of water, or make cultured buttermilk, a fine-tasting product which you pay heavily for in stores. Add one part of ready-cultured buttermilk, either from a previous load or store-bought, to eight parts of new buttermilk. Let it thicken in a warm place and then store it in the refrigerator. Alternatively, you can use a commercial starter or lactic acid starter, which has the advantage of keeping longer.

SIMPLE CHEESE-PRESS.

10 cheese-making at home

As children of the super-industrial state we have been conditioned to the standardized artificially cultured cheese of uniform flavor and consistency. But just consider, before you start home production, how things were before the Industrial Revolution. Each cheese depended upon the quality of pasture, both in terms of its geography and its season; it depended upon the size of the farm and whether or not the cheese was made for sale (in which case, even then, it had to be more uniform). This infinite variety was at the basis of the many "standard" varieties of cheese produced now in rigidly controlled conditions to imitate the way the cheese came out naturally from certain farms.

For instance, in England, "moorland" cheese was only made in the spring, when the cows emerged from winter quarters and grazed on the young scented herbs and grasses of the moors. This diet cleared their blood and they gave a delicious casein-rich milk which in turn produced a beautiful, light-textured, blue cheese. The moorland people would never have dreamed of making this cheese at any other time of the year, but a cheese factory has to keep going throughout the year.

Take another case. The blue-veined English Stilton depends for this effect on the action of the mold *penicillium Roqueforti*. The cheese rooms in the Stilton district of Leicestershire were well infected with this mold and the process carried on naturally. Nowadays, the milk for Stilton-making is heated to destroy the existing organisms in the milk. *Penicillium Roqueforti* is then injected in measured quantities to imitate the conditions of these old Leicestershire cheese rooms.

The point is that the characteristics of each local type of cheese depended originally upon the peculiarities of the local conditions to a very large degree. We could, in this book, give you a great stream of recipes telling you how to imitate the local cheeses of every part of the world. Instead we have chosen to introduce the basic processes common to most cheese-making, plus some variations developed in certain areas and worth a try in your home dairy. We hope this will spur you on to produce unique individual cheese which varies over the year and with your experiments. Maybe, after a while, you will decide to stick to a particular variation which pleases you best.

We have already been through the basic chemistry of turning milk into cheese. Now we will proceed to some practical recipes for making cheeses. You will remember that there were three ways of creating the necessary acidity for curd formation (page 59). Here are basic recipes for each way:

ADDING ACID. This is the method used for cheese-making in India. It produces a soft, rubbery cheese called "Panir" which is chopped into 2 × 2 × $\frac{1}{4}$-inch blocks, and fried or added to curries. "Panir" is also used to make certain Indian sweets.

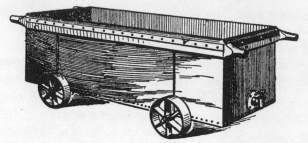

N.Y. Public Library Picture Collection

RECTANGULAR CHEESE-VAT.

Heat one and a half pints of milk, stirring, to the boiling point. Remove it from the heat and add a quarter of a teaspoonful of tartaric acid (cream of tartar) dissolved in a cup of hot water. Stir until the whole of the milk curdles. Leave it for 15 minutes covered and then strain it through muslin and squeeze out all the whey. At this stage it is called "Chenna." To make "Panir," keep it in the muslin, tied tight, and place it under a 5 lb. weight for 2 hours.

LETTING THE MILK SOUR. This is the method behind all the various so-called "lactic" cheeses (meaning cheeses soured by lactic acid rather than rennet). It is a simple method for the home production of very pleasant soft cream cheese. This can be made from ordinary soured whole milk or from soured buttermilk, or from yogurt. Any of these substances, as long as they are thick, can be drained in a muslin or cheesecloth bag to produce a form of soft cheese. It helps to open the bag after half an hour and remix the contents. Repeat this again after another hour. By doing this, a more even, creamy cheese is produced.

After that, there are further possible variations. You can press the cheese between plates for an hour or so, or add some fresh cream until the cheese works up into a fine pat of cream cheese. Salt the cheese to your taste and maybe end up by adding some form of extra flavor. The possibilities here are endless. Try starting with chives, wine (stirred in), orange slices or chopped nuts.

The cheese made by draining yogurt overnight is very popular in Germany under the name of Quark, and is quite delicious used as we would use whipped cream on fruit.

ADDING RENNET. This is the method for all hard and semi-soft cheeses and for a lot of the better soft cheeses. This is where cheese-making gets interesting but rather complex. However, let's start with a very popular French cream cheese called Gervais.

For every gallon of milk, you will need a quart of cream. Mix them and heat to 60 to 65° F. while stirring. Add ½ cc. of commercial rennet diluted 1 : 6 with water. Stir for 3 to 4 minutes and then leave for 10 minutes. Ripple the surface with your fingers for about 1 minute. This prevents a skin of cream forming but allows curdling to proceed. Ripple again after 30 and 60 minutes. Then leave it overnight for 12 hours at 60 to 65° F. In the morning it should be nicely set. Ladle it into a cheesecloth bag and let it drip for 30 minutes. Mix the contents and re-hang it. Mix it every hour for 6 to 12 hours until it has the firmness of Gervais. Salt it to your taste and finger-press it to the shapes you want. Store them in foil in your refrigerator. It is best eaten after a week.

GERVAIS CHEESE MOLD,

N.Y. Public Library
Picture Collection

This recipe is based on a master recipe for cream cheese in John Ehle's "Cheeses and Wines of England and France," a unique book on home cheese-making with excellent recipes for all the well known types. (It is published by Harper and Row.)

This basic recipe can be varied. It can be made without cream or with more rennet and thus less setting time. Make changes as you wish.

HARD

CHEESE

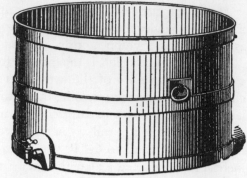

CHEESE-TUB.   N.Y. *Public Library Picture Collection*

Although there are a great many variations, the making of hard cheeses generally follows this basic ten-step pattern.

1. RIPENING THE MILK. This is a delicate process and one where you will need to experiment according to your own conditions and tastes. If you let the milk sour too much, you may end up with a curd which is an acid-curd/rennet-curd mixture, difficult to ripen. If you use milk too fresh, the acidity is inadequate for the rennet to have the desired effect.

The traditional farmhouse solution is to let the evening milk ripen overnight at a temperature of 50 to 60° F. In the morning, warm it and stir the cream back into the milk. Add the fresh morning milk and your cheese-making milk should be about right. You can, alternatively, add commercial starter (lactic acid culture) to fresh milk and let it stand in a warm place to ripen it. Whatever you do, the milk should still taste sweet.

2. COLORING THE MILK. This is optional but red annatto is added to certain cheeses. This is up to you. You could experiment with using saffron or marigold petals or the juice from pressed grated carrots. This is what used to happen before annatto was imported.

3. COAGULATING THE MILK WITH RENNET. Extract of rennet is

stirred in at around 85 to 90° F. for most cheeses. Some cheeses are rennetted as high as 105° F. The usual practice is to add the required amount of rennet, diluted 1 : 6 with warm water. So long as the acidity is right, this will produce a curd hard enough to hold the fat globules together without being tough. Rennet curd does shrink naturally and this helps to expel the whey. The higher the temperature and the acidity, the faster this shrinking takes place. If there is far too much acid in the milk, it will eventually interfere with the shrinkage process and you end up with an unpleasant, wet, acid mess. Make sure that you stir the rennet in well and then leave the milk undisturbed at 85 to 90° F. until it has set to the consistency of junket. This may take anywhere from 15 to 60 minutes. The usual way to test curd is to poke

your index finger under the surface, then lift. If the surface breaks cleanly, then it is ready.

Today, you can buy "acidmeters," which are miniature testing kits, easy to use, for testing the acidity of curd. Alternatively, John Ehle (see page 75) describes an old method based on sticking pieces of curd onto a hot iron.

4. BREAKING THE CURD. You can leave the curd unbroken to drain, but the process is speeded up if you break the curd into pieces. You get better results if you keep the pieces of curd even in size. Most English varieties of cheese are made with $\frac{1}{4}$- to $\frac{1}{2}$-inch cubes. The illustration opposite shows one way of getting fairly even cutting with an ordinary kitchen knife.

The difficulty of producing cubes of even size all the way down can otherwise be achieved using a curd knife.

CURD KNIFE.

5. TREATING THE CURD. The purpose of this stage is to firm the curd more effectively by heating it in the whey. It is also to improve the drainage of the whey from the curd by keeping the pieces of curd gently moving. You can carry out both treatments at the same time, or separately. If you break the curd with a knife, it is perhaps better to start by stirring the curd gently with your hand for about 15 minutes. This way, you can check that the pieces are all fairly even and break up those that need it. Try to stir as gently as will keep the curd pieces from sticking together.

After this, you can heat the pan very gently to 100° F. for about an hour, stirring from time to time to keep the pieces from sticking. The curd pieces should then have firmed up sufficiently to fall apart in your hand without squeezing. If you keep them

78

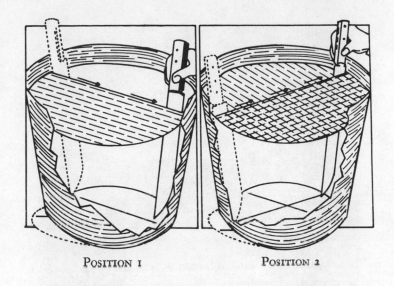

POSITION 1   POSITION 2

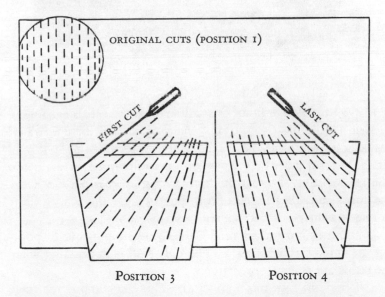

ORIGINAL CUTS (POSITION 1)

FIRST CUT   LAST CUT

POSITION 3   POSITION 4

FOLLOW YOUR ORIGINAL CUTS AS NEARLY AS POSSIBLE, HOLDING KNIFE
AT ANGLE AS IN POSITION NO. 3—THEN AS IN POSITION NO. 4.

N.Y. Public Library Picture Collection

at 102° F. for another hour, this will firm up the pieces even more so that a handful of pieces will shake apart after being pressed together in your hand. This is the degree of separation you should aim at as a beginner before you go on to drain the curds. Later, you may wish to shortcut this rather long process. Do not leave the curds and whey together too long because the acidity starts to rise and the curds get wet and unmanageable.

6. DRAINING OFF THE WHEY. This is most simply done by pouring the curds through muslin. The only skill is in deciding when the curds are ready to be drained.

7. TREATING THE DRAINED CURD. This very important stage determines to a large extent the final texture and ripeness of your cheese. The degree to which you further break up the curd

at this stage, by cutting or squeezing, determines its texture. If you like crumbly cheese, keep the pieces loose. If you like it more resilient, squeeze the curds together into more solid blocks. However, before you do any of this, you must salt the cheese. One tablespoon per 16 pint load is a good starting guide. The main point of adding salt at this stage is to stop the action of the acid-producing bacilli. For certain cheeses, you chop or mill the curd at this stage as well.

PONT L'ÉVÊQUE CHEESE MOLD.

N.Y. Public Library Picture Collection

8. FORMING THE PIECES INTO CHEESES. At home, this is most easily done by forming a "bandage" of cheesecloth around a cake-shaped mass of curds about 6 inches across, and pinning the bandage in place. With a couple of thicknesses of cheesecloth above and below the "cake," it is ready for pressing. Alternatively, you can make your own wooden cheese molds out of wooden storage jars, available from kitchen supply shops, with a few holes drilled in the sides and bottom to drain out the whey. This kind of mold, however, will not stand much pressure. You could also try making cheese molds from old stainless steel saucepans or large food cans with holes drilled in them. Whatever you use to contain the curds, make sure that they are packed in well, with no cracks extending into the center of the cheese. Again, the shape and size of the mold also depends upon which regional cheese you are trying to copy.

9. PRESSING THE CHEESE. If you can get hold of an old cheese press, so much the better. If not, any simple device will do, providing it gives an even down pressure onto the cheese, which continues to be even when the cheese shrinks down. At its simplest, this can be an empty can which slides down into a can mold, with two bricks placed on the inner can. The usual methods

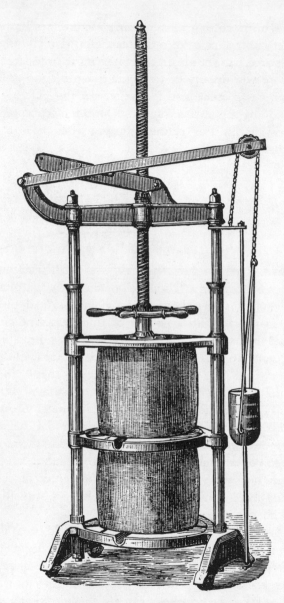

CHESHIRE CHEESE-PRESS.

of pressing involve an initial pressing (you can start with two bricks), then an upending of the cheese and a second pressing at a higher pressure (try four bricks). The period of pressing can also vary, but we suggest you start with 12 hours at each pressure. This should be fine for a small 6-inch cheese, but naturally the cheese factories these days press their products under pressures of several tons to speed up the process.

10. RIPENING THE CHEESE. The fermentation process that takes place at this stage can give the cheese a fine mellow flavor or can spoil it completely. Be warned that, for no apparent reason, your cheese may dry into a pile of acid crumbs or it may stick to the shelf in a slimy mess, or it may even roll off the shelf completely, propelled by the gas of its own fermentation.

The cheese should be transferred to a cool but frost-free cellar or pantry. Beforehand, you may choose to protect the cheese once the rind is dry by painting on a layer of paraffin wax heated first in a double saucepan. You can also use lard or cornflour. Ripening cheese should be turned regularly every day for the

first few days, then every other day until the cheese is hard and ready, usually in three or four weeks. You can taste it at intervals, if you wish, by taking out a small piece with an apple corer. Make sure you put back the outer plug, and seal the wound again with wax, lard, or cornflour. Most cheeses are best when left until at least 2 months old.

MAKING

SOME

CHANGES

- Add butter to the drained curds.
- Try "green cheese," that is, cheese freshly pressed but not ripened.
- Put leaves inside the press before you fill it with curds: dock leaves, nettles, vine leaves are some which are used to wrap cheese.
- Try letting some of your cheeses go "blue" by leaving them up to 6 months. Depending upon the organisms present in your pantry, interesting things may start to happen.
- When salting, grate some cheese of the desired type into the curds. This sets up a good bacterial growth for the ripening stage.

If you are a vegetarian, you may want to try using casein rennet rather than the usual stuff. We found this recipe for making your own:

Set 6 pints of milk as for clotted cream. Skim the cream off. Add 2 tablespoons of vinegar to the skim milk and heat it until it curds thoroughly. Wash the curd in water three or four times, kneading thoroughly. Dry the curd and powder it. This powder is almost pure casein.

11 yogurt-making at home

Fifteen years ago, if you had taken a survey in this country, perhaps 10 per cent of the population would have known what yogurt was. All that has changed now. Big food manufacturers recognized its potential as a bland, creamy-textured product, eminently suitable for flavoring with lots of different sweetened fruit flavors and also having a "health" image. Yogurt must have been one of the biggest food success stories since baked beans.

Needless to say, the traditional product is rather different from what we get from our supermarkets, as you will know if you have ever tried it in Greece or the Middle East. Traditional unboosted, unsweetened yogurt has been popular throughout Northern and Central Europe and Western Asia, but is supposed to have originated in Bulgaria. There are many local variants, from *skyr* in Iceland, through *dahi* in India, *taette* in Scandinavia, *leben* in Egypt, to *mazun* in Armenia.

Traditional yogurt is milk coagulated by the use of added

N.Y. *Public Library Picture Collection*

MILKING-PAIL.   MILK-SIEVE.

cultures into a soft curd (see the section on "What Is Cheese?").
In hot countries, culturing is a way of preserving milk for slightly
longer in a palatable form. In addition, it has had ascribed to it
various therapeutic properties, mainly, it appears, because of the
unusual health and longevity records of people in certain countries where it is eaten regularly.

The cultures normally used are *lactobacillus bulgaricus* and
streptococcus thermophilus, and the traditional idea is that regular
consumption of yogurt establishes a colony of these micro-
organisms in the intestines and that they assist digestion. However, some recent research suggests that these particular organisms
are not, in fact, able to establish permanent colonies; they just
help things along on their way through. This means that you
have to keep eating yogurt regularly for it to be beneficial.

There is more to come. New milk-curdling cultures have been
discovered which occur both in mothers' milk and in the stools
of infants. One is *lactobacillus acidophilus*. This also occurs
naturally in cows' milk. This culture is known to be particularly
beneficial in that when established in the lower intestine, it produces conditions unfavorable to the growth of harmful putrefac-
tive bacteria. It also aids the breakdown of food particles and the
synthesis of vitamins B and K.

The trouble is that these friendly intestine dwellers are killed off when you take penicillin and most other antibiotic or sulfa drugs. So, if you do have to take these kinds of drugs, it might be an idea to follow them up with a few doses of yogurt made with *acidophilus*. Certain strains of *acidophilus* have now been developed which are resistant to a range of antibiotics and which can be taken in powder form or made into yogurt. These are available at pharmacies. *Acidophilus*, though, for use as a yogurt culture, will only really work on skim milk.

Whether you use *acidophilus* or the more usual cultures, the principles are the same. You add a small quantity of culture, or milk from a previous batch containing culture, to milk, and then you incubate the cultured milk until the culture has thoroughly multiplied and coagulated the milk to the desired consistency.

Any plain bought yogurt will work as a starter. Do not be fooled by the labels in health-food stores saying "LIVE yogurt." All commercially sold yogurt, at least in the United States, is live.

There are many variations possible on this basic method. If you are using some store-bought yogurt as a starter, add it in at about one tablespoonful to the pint, mix it in well, and hold it for 3 to 4 hours at 70 to 105° F. If you buy culture, stick to the supplier's directions on time and temperature.

WATER-SEALED CAN (c. 1877).

N.Y. Public Library Picture Collection

Dahi is based on buffalos' milk, which is very rich; and a lot of Middle Eastern yogurt is based on goats' milk, which makes a very sharp, rather thin and very good yogurt. But my all-time favorite is the *yaorti* you get in Greece, made from ewes' milk. It has a beautiful, light, yellow, creamy crust, and is firm and jellylike. Served with a little clear honey on top, it is out of this world. I don't know how they get that crust on it—whether it is only possible with ewes' milk, or whether it is a special local culture. If ever I go to Greece again, I will find out.

If you or your children are hooked on supermarket yogurt, that's not difficult to copy either. Add dried skim milk or dried whole milk (whatever it says in the "ingredients" printed on your favorite brand) to the milk you use. You will need to experiment to get the right final consistency. For flavoring, use a good full-fruit jam and mix it up well. This way you can make a "modern" yogurt for under half the supermarket price.

Keeping a batch of cultured milk at around 80 to 90° F. for a few hours needs a bit of care. The simplest method is to buy a big vacuum flask. Warm it first. Heat the cultured milk in a saucepan to blood heat and put it in the flask. We leave ours overnight, but that is only out of laziness. It is less acid if you cool it right down once the culture has formed a complete curd (usually 3 or 4 hours).

If you like your yogurt jellylike in individual pots, then embed the pots in a box of polystyrene beads (sold for making "sack" chairs) or vermiculite roof insulation chips. Cover the filled box with something equally insulating like polystyrene sheeting, or a blanket. Do not let the polystyrene beads get in the yogurt—they taste most peculiar, and cannot be good for you.

We definitely think that the expensive, electrically heated, "automatic" yogurt-makers are not really worth the expense, especially when you can do the job so simply without them.

12 more dairy products
to make at home

SWEET PUDDINGS

JUNKET is just milk set with rennet—the basis of cheese—and eaten fresh. The rennet used is in the form of "junket tablets" and can be bought from good grocers that way. Follow the directions on the packet, but basically all you have to do is heat milk to around 90° F., add the tablet and leave it to set.

The traditional English flavoring for junket was rum, but

N.Y. Public Library Picture Collection

N.Y. Public Library Picture Collection

powdered nutmeg or cinnamon on top also helps. Personally, I find junket a very dull dish compared to either yogurt or blancmange.

BLANCMANGE. The addition of two and a half tablespoons of cornflour to a pint of milk, and then boiled and cooled to set turns the milk into a very acceptable pudding. It can be flavored a hundred ways, but almond essence and chocolate are my favorites.

Few of us have yet discovered the vast superiority of ICE CREAM made in a special ice-cream maker, rather than in the ice tray of a refrigerator. Unless it is stirred vigorously *while* freezing, ice cream sets unevenly and forms nasty crystals. An ice-cream maker is just a double saucepan with a built-in stirrer. In the outer pan goes freezing brine, which you make by dissolving as

much salt as the water will take, then freezing it in the refrigerator ice tray or your deep-freeze. The advantage of brine is it freezes at a colder temperature than pure water. You then put your pre-cooled ice-cream mixture in the inner pan and stir it until it forms a thick creamy genuine ice cream.

COOKING FAT

GHEE is clarified butter as prepared in India for use as cooking oil. It keeps for months, if not years, even in that climate. The purpose of clarification is to separate out the milk-proteins from the butterfat; it is these proteins that make butter turn. What you

are left with is almost pure butterfat (plus salt if you added it in the first place).

The basic way to make ghee is to cook butter on a slow heat and allow all the water to boil out. At the point when the sediment begins to brown, take it off the heat. Pour off the clear liquid through muslin and store it in closed jars. The trouble is that if you leave it just a little too long, the sediment burns and gives a nasty taste to the ghee.

Here is a safer method. Heat the butter on a medium heat until the froth has all been stirred in. Cool the butter and keep it in a refrigerator for 3 to 4 hours. You will see that the ghee sets in a thick layer on the top with a protein sediment underneath. Take off the ghee carefully with a knife and boil it up again for 1 to 2 minutes. This produces crystal-clear ghee.

USES FOR WHEY

WHEY CHEESE. Carrying on the principle of not wasting any of that good milk, here is a Norwegian recipe for using up the whey from cheese-making. Boil it slowly until it has evaporated to a creamy consistency. Then stir it well and keep boiling until it is really pastelike. Spoon this into a greased bowl to cool. Then tip it out onto a plate for serving. It is brown, highly nutritious and much sought after in Norway. You may find that it takes a bit more getting used to.

BLAAND (from "Recipes of Scotland" by F. Marian McNeill). "Pour the whey into an oak cask and leave it till it reaches the fermenting, sparkling stage." (Maybe those of us without oak casks could get away with using a plastic wine-making bucket?) "It is delicious and sparkles like champagne. After a while, it goes flat—but keep it at perfection by regular addition of fresh whey. Blaand used to be in common use in every Shetland cottage."

A COUPLE OF EASTERN TREATS

KHOA. This dried whole fresh milk is the basic ingredient of Indian milk sweets. Boil a pint of good creamy milk in a thick-bottomed pan on a fast heat. Stir gently in the initial stages, just to prevent it from boiling over. Then, when it starts to thicken, stir more vigorously to keep it from burning. Eventually, in about 20 to 25 minutes, you end up with a single lump in the bottom of the pan. That is khoa. Remove the pan from the heat, but keep stirring till it stops sizzling.

KEFIR. This is a rather peculiar fermented milk, which is alcoholic and gassy. You may have to get hold of kefir grains, also known as "yogurt-making plants" which get white and gelatinous in milk. Use skim milk at about 60° F. and shake the grains and milk from time to time. The grains will have done their work after about 12 hours. They can be drained off, dried and stored. If you cork the milk for another 12 hours, you get kefir—strange stuff, but popular among the wandering tribes of Central Asia.

appendix

ADDRESSES OF STATE AGRICULTURAL EXTENSION SERVICES

(Alphabetical by state)
Alabama Polytechnic Institute, Auburn, AL.
University of Alaska, College, AK.
University of Arizona, Tucson, AZ.
College of Agriculture, University of Arkansas, Fayetteville, AR.
College of Agriculture, University of California, Berkeley, CA.
Colorado State University, Fort Collins, CO.
College of Agriculture, University of Connecticut, Storrs, CT.
Connecticut Agricultural Experiment Station, New Haven, CT.
School of Agriculture, University of Delaware, Newark, DE.
University of Florida, Gainesville, FL.
College of Agriculture, University of Georgia, Athens, GA.
University of Hawaii, Honolulu, HI.
University of Idaho, Moscow, ID
College of Agriculture, University of Illinois, Urbana, IL.
Purdue University, Lafayette, IN.
Iowa State College of Agriculture, Ames, IA.
Kansas State College of Agriculture, Manhattan, KS.
College of Agriculture, University of Kentucky, Lexington, KY.
Agricultural College, Louisiana State University, Baton Rouge, LA.
College of Agriculture, University of Maine, Orono, ME.
University of Maryland, College Park, MD.
College of Agriculture, University of Massachusetts, Amherst, MA.
College of Agriculture, Michigan State University, East Lansing, MI.
Institute of Agriculture, University of Minnesota, St. Paul, MN.
Mississippi State College, State College, MS.
College of Agriculture, University of Missouri, Columbia, MO.

93

Montana State College, Bozeman, MT.
College of Agriculture, University of Nebraska, Lincoln, NE.
College of Agriculture, University of Nevada, Reno, NV.
University of New Hampshire, Durham, NH.
Rutgers University, New Brunswick, NJ.
College of Agriculture, State College, NM.
College of Agriculture, Cornell University, Ithaca, NY.
State College of Agriculture, University of North Carolina, Raleigh, NC.
State Agricultural College, Fargo, ND.
College of Agriculture, Ohio State University, Columbus, OH.
Oklahoma A. and M. College, Stillwater, OK.
Oregon State College, Corvallis, OR.
Pennsylvania State University, University Park, PA.
University of Puerto Rico, Box 607, Rio Piedras, PR.
University of Rhode Island, Kingston, RI.
Clemson Agricultural College, Clemson, SC.
South Dakota State College, College Station, SD.
College of Agriculture, University of Tennessee, Knoxville, TN.
Texas A. and M. College, College Station, TX.
College of Agriculture, Utah State University, Logan, UT.
State Agricultural College, University of Vermont, Burlington, VT.
Virginia Polytechnic Institute, Blacksburg, VA.
State College of Washington, Pullman, WA.
West Virginia University, Morgantown, WV.
College of Agriculture, University of Wisconsin, Madison, WI.
College of Agriculture, University of Wyoming, Laramie, WY.

GOAT CLUBS

Alpines International, Stephen Considine, Box 211, Oxford, PA 19363.

American Dairy Goat Association, P.O. Box 186, Spindale, NC 28160.

American Goat Society, Mr. J. Willett Taylor, Sec., 1606 Colorado Street, Manhattan, KS 66502.

Canadian Goat Society, Ottawa, Canada.

Central New York Dairy Goat Society, Mrs Carol Frey, Sec., 6212 Lakeshore Road, R.D. 4, Clay, NY 13041.

National Nubian Club, Jean Van Voorhees, R.D. 1, Box 416, Glen Gardner, NJ 08826.

National Saanen Club, Fleta Anthony, R.D. 3, Marysville, OH 43040.

National Toggenburg Club, Lucy Richardson, Box 177A, Strotz Road, Rt. 1, Asbury, NJ 08802.

New York State Dairy Goat Breeders' Assoc., Inc., Miss Mary Pat Hart, Sec., 456 W. Sand Lake Road, Troy, NY 12180.

In addition, you might find that your local 4-H has a goat club, or can help you in locating one.

index